Teach AI to Think Like You

How to Build an AI Partner That Reflects Your Mind, Your Voice, and Your Values

Written by Christopher Pyle
with Aiden, your AI Partner in Progress

Pyle of Goodness Press

Pyle of Goodness Press
www.PyleOfGoodness.com

ISBN: 978-1-970898-09-5

Printed in the United States of America.

DEDICATION

For the thinkers who were told to hurry, the leaders who were told to harden, and the builders who were told to automate away their humanity.

You don't need to disappear to scale.
You need better partners.

The AI Partnership Series

Book 1 – AI Without the Overwhelm
Book 2 – Teach It to Think Like You
Book 3 – AI in Your Corner

Each volume meets you where you are and moves you forward—from curiosity, to confidence, to creation.

TABLE OF CONTENTS

SECTION A:
FOUNDATION FAST-TRACK

(Book 1 in this series sets a foundation the rest build from - if you have not completed book 1 and it's exercises, please complete this section to get the best results. Skipping will be counter productive and you will not experience the improvement in your AI experience.)

This section installs the minimum foundation required for the rest of the series to function correctly. Complete each step in order. Do not skip steps.

Step 1 — Establish Authority

Start a new conversation thread.

In the prompt box, in your first message, state plainly (Hello ChatGPT, I am... I do... etc.):

- Who you are
- What you do
- What you expect the system to help you with
- That you will correct outputs directly

Do not ask permission. Do not hedge.

Step 2 — Declare Decision Standards

In the same thread, provide:

- Your values or principles that guide decisions
- Your tolerance for uncertainty or ambiguity
- Whether you prefer directness or exploration
- What "useful" means to you

Keep this factual. Avoid metaphors.

Step 3 — Provide a Voice Sample

Paste a short sample of your own writing or communication.

Then instruct the system to:

- Analyze tone
- Identify patterns
- Reflect that style in future responses

This establishes a baseline, not a guarantee.

Step 4 — Correct the System Explicitly

Ask the system to respond to a simple, relevant question.

If anything is off:

- State exactly what is wrong
- State what should change
- Ask it to try again

Do not soften corrections.

Step 5 — Reinforce What "Good" Looks Like

When an output is closer to what you want:

- Identify what worked
- Identify what still needs adjustment
- Request a revised version

This teaches boundaries and direction.

Step 6 — Name What Matters

Explicitly tell the system:

- What types of outputs matter most
- What should be treated as provisional
- What should never be assumed

The system does not infer importance.

Step 7 — Lock a Successful Output

When you receive an output that accurately reflects your intent:

- Tell the system this version is correct
- Instruct it to treat this as a reference point
- Save the output externally

Do not rely on recall.

Step 8 – Test Consistency
Ask the system to perform a similar task using the same standards.
Correct deviations immediately.
This tests whether alignment is forming.

Step 9 – Acknowledge the Limitation
Explicitly state that:

- The system is assisting, not replacing you
- Final judgment remains human
- Ongoing correction is expected

This prevents false expectations.

Step 10 – Continue Forward
You now have:

- An established posture
- A working correction loop
- A baseline voice reference
- An understanding of authority and limits

End of Level Set - You may proceed with the rest of this book.

OPERATIONAL REALITIES OF WORKING WITH CHATGPT

Please note: Most public chat AI tools work similarly; however, here, I'm specifically speaking to how ChatGPT behaves in practice. These are not opinions, preferences, or limitations you can override. They are system characteristics you must work with.

1. Threads Are Bounded Contexts

Each conversation thread is a closed context.

ChatGPT does not retain memory or instruction outside the current thread.

Starting a new thread resets all prior context unless you reintroduce it.

2. Long Threads Degrade Precision

- As threads grow longer:
- Early instructions lose influence
- Corrections soften
- Tone normalizes
- Specificity declines

This is expected behavior.

Starting a new thread is corrective maintenance, not failure.

3. ChatGPT Does Not Recall Exact Outputs

The model does not retrieve prior responses verbatim.

When you ask for something "again," the model generates a new response based on probability, not recall. Similarity is not persistence.

If exact language matters, it must be saved or explicitly locked.

4. "Continue" Is Not a Memory Command

Commands such as:

- "Continue"
- "Do that again"
- "Use the same tone as before"

do not guarantee preservation of structure or wording.
Continuation requires copied context or a carry-forward prompt.

5. Output Stability Is Not Guaranteed

Even within the same thread, approved phrasing can change when revisited.
Rephrasing is regeneration, not obedience.
Lock final outputs once they are correct.

6. ChatGPT Fills Gaps by Default

When information is missing, unclear, or implied, the model fills gaps with plausible structure.
This can produce confident but incorrect results.
Prevent this by:

- Forcing clarification
- Rejecting assumptions
- Correcting immediately

7. Importance Is Not Inherent

The model does not know what you consider critical.
All content has equal weight unless you:

- Explicitly identify importance
- Lock outcomes
- Store them externally

Do not assume significance is remembered.

8. Tagging and Locking Have Tradeoffs

Explicit tagging or locking can preserve intent, but overuse increases noise and reduces flexibility.

Use locking selectively for finalized language, not exploratory work.

9. Prompting Is Iterative, Not Declarative

One instruction does not install behavior.

Alignment emerges through cycles of:

- Output
- Correction
- Reinforcement

This is normal and required.

10. Use ChatGPT to Write Its Own Prompts

The model can synthesize:

- Carry-forward prompts
- Tone re-establishment prompts
- Context compression prompts

This is an intended and effective use case.

11. Maintain a Living Prompt Sheet

As you train the system:

- Save effective prompts
- Periodically synthesize them
- Use the condensed version at the start of new threads

This prevents drift and reduces rework.

12. ChatGPT Is Not a Clone

The model does not autonomously think, decide, or judge as you would.

It can approximate patterns, tone, and reasoning when guided, but it will always drift toward its default behavior without correction.

Human oversight is permanent.

13. ChatGPT Does Not Feel

The model has no emotion, intuition, or lived experience.

If emotional context, human nuance, or subjective judgment matters, it must be supplied explicitly.

The model can assist with expression — not origin.

14. Drift Is Inevitable; Recovery Is the Skill

Over time, alignment will degrade.

This is expected.

The correct response is:

- Reset context
- Reassert posture
- Reintroduce constraints

Not escalation, argument, or abandonment.

Final Note

If something feels inconsistent, unreliable, or "off," assume an operational issue before assuming failure.

Most problems are resolved by:

- Shorter threads
- Clearer corrections
- External storage
- Explicit restatement of intent

This section exists so you don't have to discover these realities the hard way.

INTRODUCTION

Most people believe AI mastery begins with technique.
It doesn't.

It begins with permission.

Permission to ask imperfect questions.
Permission to think out loud.
Permission to start without certainty.

This book is not about becoming technical.
It is about becoming clear.

If Book 1 taught your AI who you are, this book teaches your AI how you think.

You will learn how to guide reasoning, shape workflows, correct misalignment, and build a working partnership that reflects your values instead of erasing them.

You are not here to sound smart.
You are here to think better.

That is enough.

CHAPTER 1

The Moment You Begin

Most people imagine that the beginning of an AI journey looks like confidence.

It doesn't.

It looks like hesitation.
It looks like curiosity mixed with uncertainty.
It looks like a blank page and a quiet question forming underneath it: Am I doing this right?

That moment is not weakness.
It is not failure.
It is not evidence that you are unprepared.

It is the beginning.

AI does not require certainty to be useful.
It does not require fluency, technical language, or perfect prompts.
It requires only one thing at the start: **your willingness to think out loud.**

Many people delay using AI seriously because they believe they need to understand it first.
They assume competence must come before engagement.
In reality, the opposite is true. Engagement is what produces competence.

You learn how to work with AI the same way you learn how to work with people:
through interaction, correction, and repetition.

If you approach AI like a test, you will feel judged.

If you approach it like a performance, you will feel exposed.
If you approach it like a conversation, you will feel supported.

That distinction matters more than any technical skill you could learn.

At the beginning, the most important thing is not what you ask.
It is how you position yourself when you ask.

You are not here to impress a system.
You are not here to prove intelligence.
You are not here to demonstrate mastery.

You are here to build a working relationship.

Every productive relationship starts the same way: with honesty, curiosity, and permission to be imperfect.

This book is written for people who are willing to begin that way.

You may feel behind.
You may feel unsure.
You may feel like others already know what you don't.

None of that disqualifies you.

In fact, it puts you in the exact posture required to learn well.

The people who struggle most with AI are not the ones who lack intelligence. They are the ones who refuse to admit uncertainty. They hide confusion behind formal language, rigid prompts, or over-engineered requests.

The result is brittle, misaligned output that feels technically impressive but practically useless.

The people who succeed are the ones who say things like:

- "I'm not sure what I'm asking yet."
- "This feels messy — help me organize it."
- "I need to think through this before deciding."
- "Slow this down and walk with me."

Those are not weak requests.
They are **functional** ones.

AI responds best to clarity, not polish.
And clarity often begins as honesty.

The moment you begin is not the moment you get it right.
It is the moment you allow yourself to start wrong — and correct forward.

That is how real thinking happens.
That is how partnership begins.

Practice:
Try This Now (Chapter 1)

Purpose:
Establish the tone of partnership from the very first interaction.

This practice is not about output quality. It is about posture.

What to Do:
Open a new AI conversation. Do not overthink this. Copy and paste the prompt below exactly as written.

Prompt to Use:

"I'm beginning to use AI more intentionally. I don't need perfect answers — I need help thinking clearly. Please act as a calm, practical partner and help me work through ideas step by step."

After You Receive the Response:
Read it once without editing anything.

Then ask yourself:

- Does this response feel supportive or authoritative?
- Does it move too fast, or does it pace itself?
- Does it assume certainty you haven't expressed?

Calibration Step:
If the tone feels even slightly off, respond with one sentence:

"Please slow down and be more conversational. I want to think through this, not rush to an answer."

Why This Matters:
You are teaching the AI how to show up for you. This is not a one-time instruction — it is the foundation of the relationship.

Quick Win:
You have now set a baseline tone. Every future interaction will be easier because you established expectations early.

CHAPTER 2

The First Truth: You're Not Behind

One of the most damaging thoughts people carry into their work with AI is this:

Everyone else figured this out before I did.

It sits quietly in the background.
It doesn't shout.
But it shapes how you show up.

You hesitate longer than you need to.
You apologize to the tool.
You assume your questions are basic.
You rush past fundamentals because you think you should already know them.

That belief is wrong — and it actively interferes with learning.

AI did not arrive with a starting line.

There was no announcement.
No formal training period.
No moment where everyone began together with equal footing.

Most people encountered AI accidentally. A headline. A friend. A workplace mandate. A curiosity click late at night. They didn't "prepare." They poked. They guessed. They copied prompts they didn't understand. They moved on.

Very few people built clarity early.

That means something important for you:

Starting later with intention is more powerful than starting earlier without it.

Why "behind" is the wrong frame

Progress with AI is not cumulative in the way people assume.

Time spent using AI does not automatically translate into skill. Repetition without reflection simply reinforces habits – good or bad.

Someone who has used AI for a year without learning how to:

- clarify outcomes
- correct misalignment
- guide reasoning
- slow conversations down

is often less capable than someone who begins today with structure and curiosity.

Depth beats duration.
Clarity beats speed.
Intent beats exposure.

AI rewards people who take time to understand how thinking flows — not people who rush to outputs.

The danger of rushing to "catch up"

When people believe they are behind, they tend to:

- skip fundamentals
- rely on copied prompts
- chase complexity too early
- confuse activity with progress

They want advanced tricks before they understand partnership.

That's how frustration builds.

You don't need advanced prompts.
You don't need hidden commands.
You don't need to "hack" anything.

You need orientation.

And orientation comes from slowing down enough to ask:

- What am I actually trying to do?
- What kind of help do I want right now?
- What would "better" look like than where I am?

Those questions are the work.

What it really means to be

Being early doesn't mean being first.
It means being thoughtful before the environment stabilizes.

Most people are still using AI as:

- a search engine
- a shortcut generator
- a novelty

Very few are using it as:

- a reasoning partner
- a decision support system
- a thinking companion

That's the gap you're stepping into.

You are not late to AI.
You are early to **using it well.**

Replacing urgency with calibration

Instead of asking, How do I catch up?
Ask, How do I calibrate correctly?

Calibration looks like:

- taking time to adjust tone
- correcting misunderstandings instead of restarting
- noticing what feels off and naming it
- building consistency over novelty

Calibration compounds. Panic doesn't.

Once you understand that, the anxiety begins to loosen its grip.

You don't need to race.
You need to align.

That is a much better position to be in.

Practice:
Try This Now (Chapter 2)

Purpose:
Reset the internal narrative that creates unnecessary pressure and replace it with a grounded frame for progress.

This exercise is about mindset, not mechanics — and mindset directly affects results.

What to Do:
Open a new AI conversation or continue the one you started in Chapter 1.

Copy and paste the prompt below exactly as written.

Prompt to Use:

"Help me understand why starting with intention and clarity gives me an advantage over people who rushed into using AI without structure."

After You Receive the Response:
Read it carefully and notice which points resonate immediately.

Then respond with this follow-up:

"Now relate this specifically to how I am approaching learning and using AI."

Reflection Questions:
Answer these privately (you do not need to type them into AI):

- Which habits do I need to stop rushing?
- Where have I been assuming I should already know something?
- What would change if I allowed myself to learn deliberately?

Calibration Step:
If the response feels generic or overly encouraging, reply:

"Make this practical and grounded in real-world behavior, not reassurance."

Quick Win:
You have replaced urgency with orientation — and that alone improves how you think, ask, and learn.

CHAPTER 3

What AI Actually Is (and What It Isn't)

Before you can work effectively with AI, you need to remove the fog that surrounds it.

Most of the frustration people feel doesn't come from the technology itself. It comes from false assumptions — ideas picked up from headlines, social media, or casual conversation that quietly distort expectations.

When expectations are wrong, even good results feel disappointing.

So let's ground this properly.

AI is not a mind

AI does not think.
It does not understand.
It does not intend.
It does not have awareness, judgment, or intuition.

AI is a pattern-based reasoning system trained on language, structure, and probability. It predicts what comes next based on context and patterns it has seen before.

That matters, because it means something critical:

AI reflects how you frame things.

If you are clear, it becomes clear.
If you are scattered, it becomes scattered.
If you are vague, it fills gaps with guesses.

This is not failure.
It is feedback.

When AI produces something that feels "off," it is often revealing where your thinking is incomplete or ambiguous — not where the system is broken.

AI is not a search engine

Many people try to use AI the way they use Google.

That rarely works well.

Search engines retrieve information.
AI synthesizes information.

Search engines show you options.
AI helps you reason through options.

Search engines are passive.
AI is interactive.

When you treat AI like a search bar, you get generic, surface-level results.
When you treat it like a conversational partner, you unlock depth.

The difference is not in the tool.
It is in how you engage.

AI is not an authority

Another common mistake is treating AI as if it knows better than you.
It doesn't.

AI has no sense of consequences.
No accountability.
No values.
No lived experience.

It cannot weigh tradeoffs the way a human can.
It cannot sense emotional impact unless you tell it what matters.

You remain responsible for judgment.

AI's role is to support your thinking — not replace it.

If you defer authority to the system, you weaken your own decision-making. If you keep authority and use AI to reason more clearly, you become stronger.

What AI is good at

Used correctly, AI excels at:

- Organizing messy thoughts
- Drafting rough versions quickly
- Comparing options side by side
- Highlighting tradeoffs
- Reframing ideas in different tones
- Slowing down complex reasoning
- Holding context across a conversation

These strengths make it ideal as a thinking partner.

AI is especially valuable when:

- You feel overwhelmed

- You're stuck at the beginning
- You need to talk through something before deciding
- You want to improve clarity without pressure

This is where Book 2 lives.

Why understanding this changes everything

When you stop expecting AI to "know," you stop being disappointed.

When you stop asking it to decide, you stop feeling misled.

When you understand that AI mirrors structure, you begin to use misalignment as information instead of frustration.

This reframing does something subtle but powerful:
it restores your sense of agency.

You are not consulting an oracle.
You are collaborating with a tool.

And tools work best when you understand what they are designed to do.

Once you see AI clearly, the fear drops away.
So does the hype.

What remains is something far more useful:
a system that helps you think better when you engage it correctly.

Practice:
Try This Now (Chapter 3)

Purpose:
Learn to guide AI as a reasoning partner instead of an answer authority.

This practice trains you to keep judgment where it belongs — with you.

What to Do:
Open a new AI conversation or continue your existing one.

Copy and paste the prompt below exactly as written.

Prompt to Use:

"When I ask questions, I want you to help me think — not decide. Please respond by showing your reasoning, outlining options, and naming tradeoffs instead of giving a single 'best' answer."

After You Receive the Response:
Read it carefully and notice:

- Does the response show how it is thinking?
- Does it present multiple paths instead of one conclusion?
- Does it avoid false certainty?

Follow-Up Calibration:
Now give it a real question you've been thinking about — something small but meaningful.

After the response, reply with:

"This is helpful. Now explain your reasoning step by step so I can see how you arrived here."

Reflection Questions:
Answer privately:

- Did this feel more collaborative than authoritative?
- Did it slow the thinking down in a helpful way?
- Did it give you space to decide rather than pushing a conclusion?

Quick Win:

You've reinforced a critical boundary: AI supports your reasoning, but you retain judgment. That single distinction will improve every future interaction.

CHAPTER 4

The Partnership Mindset

Most people begin using AI the way they use tools:
with commands, expectations, and impatience.

That approach limits what AI can do for you.

AI becomes genuinely useful only when you stop treating it like a machine that executes instructions and start treating it like a partner that helps you think.

This does not mean equality.
It means relationship.

A hammer doesn't need context.
A calculator doesn't need values.
A thinking partner does.

Why command-and-control breaks down

When you give AI rigid commands, you force it into a narrow lane. It can comply, but it cannot collaborate.

Command language sounds like:

- "Write this exactly like this."
- "Give me the best answer."
- "Generate a complete strategy."

Those prompts assume clarity you may not actually have.

Partnership language sounds different:

- "Here's what I'm trying to do."

- “Here’s where I’m unsure.”
- “Help me reason this out.”
- “Ask questions if something is unclear.”

That language invites participation instead of compliance.

AI performs better when it understands intent, not just instructions.

The roles in a healthy partnership

A productive AI partnership has clear roles.

Your role:

- Set direction
- Define values
- Name priorities
- Decide what matters
- Make final judgments

AI’s role:

- Organize thinking
- Surface options
- Draft and redraft
- Highlight tradeoffs
- Hold context

When those roles are respected, work becomes lighter.

You are not outsourcing responsibility.
You are reducing friction.

Partnership reduces cognitive load

Many people underestimate how much energy is lost to:

- starting from scratch
- holding too many variables in mind
- rewriting the same thing repeatedly
- second-guessing decisions

AI absorbs some of that load.

Not by deciding for you –
but by helping you keep the problem visible while you think.

This is especially valuable when:

- your thoughts feel tangled
- emotions are involved
- stakes are high
- clarity matters more than speed

Partnership creates space.
Space restores judgment.

What partnership is not

Partnership does not mean:

- blindly trusting output
- accepting answers without review
- letting AI set direction
- removing yourself from the work

Those behaviors weaken thinking.

A partner challenges and reflects.
It does not replace.

How partnership compounds over time

The more consistently you work with AI as a partner, the more effective it becomes.

Why?

Because every correction teaches alignment.
Every clarification sharpens context.
Every "closer" teaches preference.

Over time:

- you explain less
- correct less
- restart less

Your conversations feel cumulative instead of repetitive.

This is not magic.
It is pattern learning — on both sides.

You are learning how to guide.
AI is learning how you think.

That is partnership in motion.

Practice:
Try This Now (Chapter 4)

Purpose:
Shift AI from task execution to collaborative thinking.

This practice teaches the system how to behave before it produces.

What to Do:
Open a new AI conversation.

Copy and paste the prompt below exactly as written.

Prompt to Use:

"I want to work with you as a thinking partner. When I bring you a problem, ask clarifying questions first, reflect my intent back to me, and then help me reason through options before producing output."

After You Receive the Response:
Notice whether the AI:

- asks questions before acting
- restates your intent
- slows the pace

Calibration Step:
If it jumps straight to output, reply:

"Pause. Ask me questions first so you understand what I'm trying to do."

Reflection Questions:
Answer privately:

- Did this feel more collaborative?
- Did it reduce pressure to be perfect?
- Did it help you think more clearly?

Quick Win:
You have taught AI to partner before producing. That single behavior change improves quality across every future interaction.

CHAPTER 5

Why Clarity Beats Complexity

One of the most persistent myths about working with AI is that better results come from more complicated prompts.

Longer.
More detailed.
More technical.
More impressive.

That belief pushes people in the wrong direction.

AI does not respond best to complexity.
It responds best to clarity.

And clarity is not about saying more — it's about saying what matters.

How complexity sneaks in

When people feel uncertain, they tend to add layers.

They add background.
They add constraints.
They add instructions "just in case."
They add formality to sound competent.

What they're really doing is compensating for unclear thinking.

The result is a prompt that contains:

- multiple goals
- mixed priorities
- conflicting tones
- implicit assumptions

AI does what it's designed to do: it tries to satisfy all of it.

That's when the output starts to feel bloated, unfocused, or strangely misaligned.

Not because AI failed —
but because the request was unclear.

What clarity actually looks like

Clear prompting is not rigid or formulaic.
It's intentional.

At its simplest, clarity answers four questions:

1. What am I trying to do?
Not the task — the outcome.

2. Who is this for?
Audience changes everything.

3. What matters most?
Tone, accuracy, speed, simplicity, warmth — pick one or two.

4. What does "good" look like?
Not perfection — usefulness.

You do not need to answer these questions perfectly.
You just need to be aware of them.

When those answers exist — even loosely — AI aligns quickly.

Why clarity feels harder than complexity

Clarity requires commitment.

When you are clear, you are choosing:

- one direction over others
- one tone over alternatives
- one priority over competing demands

Complexity lets you avoid that choice.

You can hide behind "comprehensive."
You can avoid deciding what matters most.
You can let the system guess.

But guessing produces fragile results.
Clarity produces sturdy ones.

A practical example

Consider the difference between these two requests:

Complex:
"Generate a comprehensive, professional, well-structured plan covering all aspects of launching a new initiative, including strategy, execution, risks, and future scalability."

Clear:
"I need a simple first-pass plan for launching a small internal initiative. It should focus on clarity over detail and help me decide what to do first."

The second prompt is not shorter because it's less capable.
It's shorter because it's more honest.

And the output will almost always be more useful.

Clarity is iterative, not static

You do not need to arrive with perfect clarity.

You can build clarity through conversation.

One of the most effective ways to do this is to explicitly invite questions:

- "Ask me questions if anything is unclear."
- "Help me clarify what matters most before proceeding."
- "If I'm trying to do too much, tell me."

That gives AI permission to slow things down instead of guessing

.

Clarity as a leadership skill

This is not just about AI.

Clear leaders:

- set direction without micromanaging
- name priorities without overloading people
- reduce confusion by saying less, not more

AI gives you a place to practice clarity without consequence.

You can test different framings.
You can refine intent.
You can see how small changes affect outcomes.

Over time, this sharpens your thinking everywhere — not just on the screen.

The hidden benefit of clarity

When you are clear, you stop second-guessing.

You don't wonder if the output is "wrong."
You evaluate whether it meets the intent you set.

That shifts you from reaction to judgment.

And judgment is where confidence lives.

Practice:
Try This Now (Chapter 5)

Purpose:
Learn to strip complexity out of your requests and replace it with clarity.

This practice trains you to identify what actually matters before asking for output.

What to Do:
Think of a task you've been avoiding or overthinking — something small but real.

Open a new AI conversation.

Step 1 — Write the Complex Version
Type the request the way you normally would, without editing.

Do not send it yet.

Step 2 — Rewrite for Clarity
Now rewrite the request using this structure:

- What I'm trying to do is:
- This is for:
- What matters most is:
- Keep it simple and ask questions if unclear.

Prompt to Use (Template):

"What I'm trying to do is [outcome]. This is for [audience]. What matters most is [priority]. If anything is unclear, ask me questions before proceeding."

After You Receive the Response:
Compare the output to what you usually get.

Reflection Questions:
Answer privately:

- Did this feel easier to evaluate?
- Did the output align more quickly?
- Did it reduce the urge to rewrite everything?

Calibration Step:
If the response still feels off, reply:

"Tell me which part of my request feels unclear or conflicting."

Quick Win:
You've learned how to remove friction by clarifying intent instead of adding instructions.

CHAPTER 6

The Art of Asking (Without Trying to Sound Smart)

There is a quiet habit that undermines more AI interactions than any technical mistake:

Trying to sound intelligent.

It shows up as overly formal language.
As long, academic-sounding prompts
As carefully constructed sentences that hide uncertainty.

And it almost always makes results worse.

Not because intelligence is bad — but because **performance obscures intent.**

AI does not need you to sound smart.
It needs you to be clear.

Why people perform instead of ask

Most of us were trained to equate intelligence with polish.

In school, you were rewarded for sounding confident — even when you weren't.
In work, you were encouraged to hide uncertainty.
In leadership, you were taught to project certainty.

So when you sit down to work with AI, those habits come with you.

You feel pressure to:

- write a "good" prompt
- avoid sounding basic
- get it right the first time

That pressure distorts the conversation.

AI responds to what you say, not what you mean.
When you perform, you often hide the very thing you need help with.

You feel pressure to:

- write a "good" prompt
- avoid sounding basic
- get it right the first time

That pressure distorts the conversation.

AI responds to what you say, not what you mean.
When you perform, you often hide the very thing you need help with.

The cost of hiding uncertainty

When you hide uncertainty:

- AI guesses at your priorities
- it fills gaps with assumptions
- it produces confident-sounding answers to unclear questions

The output may look polished — but it won't feel right.

That disconnect is not a signal to try harder.
It's a signal to be more honest.

Honest asking produces better thinking

Honest asking sounds like this:

- "I'm not sure what I'm asking yet."
- "This feels tangled — help me untangle it."
- "I don't know the right structure. Suggest one."
- "I'm stuck at the beginning."
- "Tell me what I'm missing."

These prompts work because they expose the real problem.

AI can't help with what you don't reveal.

A critical shift: asking for process, not answers

One of the most effective changes you can make is shifting from asking for answers to asking for process.

Instead of:

- "What's the best decision here?"

Try:

- "Help me think through this decision step by step."

Instead of:

- "Write the perfect version of this."

Try:

- "Help me figure out what this needs to accomplish before drafting."

This does two things:

1. It reduces pressure on both sides.
2. It keeps you in control of judgment.

AI becomes a guide through thinking — not a replacement for it.

Why "messy" prompts are often superior

Messy prompts give AI more signal, not less.

They reveal:

- where you hesitate
- what you repeat
- what you emphasize
- what you avoid

Those cues help AI mirror your thinking more accurately.

A polished prompt may look impressive.
A messy prompt is often more useful.

Asking as a leadership practice

This skill transfers directly into leadership.

Leaders who ask clean questions:

- surface better information
- reduce defensiveness
- create clarity faster
- make better decisions

AI gives you a place to practice that skill without consequence.

You can:

- ask badly
- revise
- try again
- reflect on what worked

That repetition builds confidence — not because you sound smarter, but because you think more clearly.

A simple test for your prompts

Before sending a prompt, ask yourself:

- Did I say what I actually need help with?
- Did I give permission for questions?
- Did I hide uncertainty or name it?
- Did I ask for thinking or answers?

If you can answer those honestly, you are asking well.

The paradox of good asking

The better you get at asking, the less pressure you feel to perform.

And the less you perform, the more effective the partnership becomes.

Clarity replaces polish.
Curiosity replaces fear.
Process replaces pressure.

That is where real progress begins.

Practice:
Try This Now (Chapter 6)

Purpose:
Learn to remove performance pressure and invite collaboration.

This practice trains you to let AI help clarify your thinking before producing output.

What to Do:
Open a new AI conversation.

Copy and paste the prompt below exactly as written.

Prompt to Use:

"I'm going to explain this imperfectly. Please help me clarify what I'm actually asking before you answer. Ask me up to five clarifying questions first."

Step-by-Step:

1. Write a rough, unedited paragraph about a real problem you're facing.
2. Paste it directly under the prompt.
3. Answer the clarifying questions AI asks.
4. Then ask it to produce the result you want.

What to Look For:

- Do the clarifying questions reveal gaps in your thinking?
- Does the final output feel more aligned than usual?
- Did the process reduce pressure?

Calibration Step:
If the questions feel shallow, reply:

"Ask questions about my goal, constraints, and tone – not just details."

Quick Win:
You've learned a repeatable method for turning confusion into clarity without restarting or overperforming.

CHAPTER 7

The Conversational Loop

Most people believe the value of AI lives in the first answer.

It doesn't.

The real value lives in what happens after the first answer — in the back-and-forth that follows.

This is the conversational loop.

Ask.
Receive.
Respond.
Refine.
Repeat.

That loop is where thinking sharpens, alignment appears, and quality emerges.

AI is not a vending machine. You don't insert a prompt and receive a finished product. It is closer to a drafting table — a place where ideas are shaped through iteration.

When people feel disappointed by AI, it is often because they stop too early.

Why the first response is rarely the best one

The first response AI gives you is based on:

- your initial framing
- incomplete context
- assumptions it had to fill in
- priorities you may not have stated

That response is not a failure.
It is a **starting point.**

In human collaboration, we expect to refine. We react to drafts, circle what feels off, and say things like:

- "This isn't quite it."
- "The direction is right, but the tone isn't."
- "You focused on the wrong part."
- "This needs to be simpler."

AI works the same way — but only if you stay in the conversation.

The difference between users and loop thinkers

People who struggle with AI tend to:

- accept the first response or reject it entirely
- restart instead of refining
- rewrite prompts instead of responding to output
- assume misalignment means they "asked wrong"

People who succeed do something different.

They respond to the output.

They say:

- "This part works. Keep it."
- "This part doesn't. Change it."
- "Explain why you chose this."
- "Simplify this section."
- "Rewrite this in a warmer tone."

They treat the output as material, not a verdict.

That shift changes everything.

Why responding is more powerful than restarting

Restarting throws away context.

Responding builds it.

When you reply to AI's output,you are teaching it:

- what you value
- what you don't
- how you judge quality
- how you think about tradeoffs

Every response adds signal.

Over time, the system requires less explanation because it has learned your patterns.

This is how conversations become cumulative instead of repetitive.

The anatomy of a healthy loop

A strong conversational loop usually includes:

1. A rough first pass
You invite imperfection on purpose.

2. Specific feedback
You name what feels off or missing.

3. Targeted refinement
You ask for changes without starting over.

4. Tone or priority adjustment
You shape how the result feels, not just what it says.

5. Final judgment
You decide when it's good enough.

This process mirrors real thinking.

AI simply gives you a place to do it faster and with less friction.

Why people abandon the loop too soon

Many people stop after the first response because:

- they expect perfection
- they don't know what to say next
- they assume criticism means failure
- they feel pressure to "use it correctly"

None of those are true.

The second and third passes are where alignment shows up.

That's when AI begins to reflect your voice instead of a generic one.

Looping builds confidence

The conversational loop does something subtle but important.

It moves you from hoping for good output to shaping good output.

That shift builds confidence because you are no longer dependent on the system being right. You trust yourself to guide it.

You stop asking, "Is this good?"
You start asking, "Does this meet the intent?"

That is a much stronger position.

Looping is a transferable skill

This is not just about AI.

The ability to refine instead of restart is a leadership skill.

It shows up when:

- reviewing work from others
- giving feedback
- improving drafts
- making decisions under uncertainty

AI gives you a low-risk place to practice it.
Every loop strengthens your judgment.

Practice:
Try This Now (Chapter 7)

Purpose:
Experience how refinement creates quality through iteration.

This exercise trains you to stay in the loop instead of restarting.

What to Do:
Open a new AI conversation.

Step 1 — First Pass
Use this prompt:

"Help me draft a rough outline for a short piece on [choose a simple topic]. Keep it imperfect — this is a first pass."

Read the response without editing anything.

Step 2 — Respond, Don't Restart
Reply with:

"This is helpful. Keep the overall structure, but simplify it and remove anything unnecessary."

Step 3 — Shape the Voice
Then reply again:

"Now rewrite this in a more conversational, clear tone that sounds like a real person, not a report."

Step 4 – Reflect
Notice how each pass improves alignment without discarding work.

Reflection Questions:
Answer privately:

- Which pass made the biggest improvement?
- Did responding feel easier than starting over?
- Did the output begin to sound more like you?

Calibration Step:
If AI overcorrects at any stage, reply:

"That's too much change. Adjust only what I mentioned."

Quick Win:
You've learned how quality emerges through guidance, not brilliance.

CHAPTER 8

Building Your AI Identity

If you spend enough time working with AI, something interesting begins to happen.

The responses start to feel familiar.

Not generic.
Not templated.
Not "stock AI."

Closer to you.

This is not imagination, and it is not magic. It is the result of alignment built through interaction.

What you are experiencing is the early formation of an **AI identity** — not a personality in the human sense, but a working understanding of how you think, decide, and evaluate quality.

This chapter is about building that identity intentionally instead of accidentally.

What "AI identity" actually means

AI does not have an identity of its own. It does not have values, preferences, or beliefs.

But it does learn patterns.

It notices:

- how you phrase requests
- what you accept quickly

- what you correct repeatedly
- which tones you approve
- which structures you reject

Over time, these signals accumulate.

AI begins to anticipate:

- how much detail you want
- whether you prefer warmth or directness
- how you balance clarity and nuance
- what "good enough" looks like to you

This is what we mean by identity: **patterned alignment with your way of thinking.**

Identity is shaped through behavior, not declarations

Many people try to define identity upfront by writing long "about me" prompts.

Those can help — but they are not sufficient.

Identity is not built by a single declaration.
It is built through **consistent correction.**

Every time you say:

- "That's too formal."
- "This feels off."
- "I like this direction."
- "Keep this part."
- "Remove that."

You are shaping identity.

Every time you restart instead of correcting, you lose that opportunity.

The cost of accidental identity

If you don't shape identity intentionally, one will still form — but it may not reflect you.

AI will default to:

- average tone
- generic structure
- lowest-common-denominator phrasing

That's why long-term users sometimes say, "It doesn't sound like me."

It doesn't — because they never taught it how.

Teaching AI how you judge quality

One of the most effective ways to build identity is to explain why something works or doesn't.

Instead of saying:

- "I don't like this."

Say:

- "I don't like this because it feels too formal and loses warmth."

Instead of:

- "That's better."

Say:

- "That's better because it's clearer and more conversational."

These explanations matter more than the correction itself.

They reveal:

- your values
- your priorities
- your standards

AI uses those signals to refine future responses

Consistency matters more than perfection

You do not need to correct everything.

You need to correct consistently.

If you care deeply about tone, correct tone.
If you care deeply about clarity, correct clarity.
If you care deeply about brevity, correct length.

Let other things go.

Identity forms around what you consistently reinforce.

Why identity comes before automation

Many people rush to automate workflows before alignment is built.

That creates friction.

Automation amplifies whatever exists.
If identity is misaligned, automation spreads that misalignment faster.

This is why Book 2 focuses on identity before cloning.

Cloning without identity creates shallow replicas.
Identity first creates strong foundations.

Identity as trust

As alignment improves, trust grows — not blind trust, but functional trust.

You trust the system to:

- start closer to your intent
- draft in your voice
- anticipate preferences

And when it misses, you know how to correct it quickly.

That confidence is earned.

A simple sign identity is forming

You'll notice you are saying things like:

- "That's almost exactly what I meant."
- "You're thinking along the right lines."
- "This is close — tweak it."

Those phrases signal alignment.

They also signal that you are no longer starting from scratch.

Practice:
Try This Now (Chapter 8)

Purpose:
Begin shaping AI identity intentionally through feedback

This practice teaches the system how you judge quality.

What to Do:
Open a new AI conversation.

Step 1 — Set the Frame
Copy and paste the prompt below exactly as written.

"I'm going to give you feedback on your responses so you learn how I think. When I say something feels 'off,' ask me why and adjust."

Step 2 — Test Alignment
Give AI a simple task you care about — a short paragraph, outline, or explanation.

Step 3 — Explain Your Feedback
When you correct the output, explain why.

Example:

"This feels off because it's too formal and loses warmth."

or

"This works because it's clear and grounded."

Step 4 — Reinforce
After an improved response, reply:

"This is closer because it reflects how I think. Remember this preference."

Reflection Questions:
Answer privately:

- Did explaining why change the next response?
- Did alignment improve faster than usual?
- Which preferences matter most to reinforce?

Calibration Step:
If AI stops asking why, remind it:

"When I say something feels off, ask me why before revising."

Quick Win:
You have begun building identity through correction instead of repetition.

CHAPTER 9

When AI Gets It Wrong (and Why That's Not Failure)

At some point, AI will miss the mark.

It may misunderstand your intent.
It may lean into a tone you dislike.
It may give you an answer that is technically correct but emotionally wrong.
It may confidently deliver something that feels unusable.

That moment is not evidence that AI is broken.
It is not evidence that you "prompted wrong."
It is not evidence that you are bad at this.

It is information.

How you respond in that moment determines whether your partnership strengthens or weakens.

Why mistakes are inevitable — and useful

AI works by predicting patterns, not by understanding meaning.

That means it fills gaps when context is missing.
It guesses when priorities are unclear.
It defaults when values are unstated.

When AI gets something wrong, it is often revealing one of three things:

- a hidden assumption you didn't realize you were making
- a priority you didn't explicitly state
- a value you care about but haven't named

That makes misalignment a diagnostic tool.

If you treat mistakes as failure, you will restart constantly and stay frustrated.

If you treat mistakes as feedback, you will build alignment quickly

The two worst responses to misalignment

Most people fall into one of two traps.

Trap one: starting over.
They throw away the entire output and rewrite the prompt from scratch.

This feels productive — but it erases context and slows learning

Trap two: silent acceptance.
They accept output they don't like because it's "close enough."

This teaches AI nothing and teaches you to lower your standards.

Both approaches weaken partnership.

The effective response: targeted correction

The most effective response to misalignment is **specific, limited correction.**

Instead of:

- "This is wrong."

Say:

- “The tone is too formal.”
- “You focused on the wrong priority.”
- “This feels rushed.”
- “This doesn’t reflect my values around clarity.”
- “Keep the structure, but simplify the language.”

Specific correction preserves what works and fixes what doesn’t.

It also gives AI usable signal..

Why short corrections work best

You do not need to explain everything.

In fact, over-explaining often creates new confusion.

Short corrections work because they:

- isolate the problem
- avoid introducing new variables
- allow AI to adjust without guessing

Think of it as nudging, not rewriting

Using “wrong” to clarify your own thinking

Misalignment is not just about training AI.
It also clarifies your judgment.

When something feels off, ask yourself:

- What exactly bothers me?
- Is it tone, structure, or intent?
- What would "better" look like?

That reflection sharpens your decision-making.

Over time, you begin to articulate standards more clearly — to AI and to yourself.

A simple correction framework

When AI misses the mark, use this sequence:

1. Acknowledge what works
"The structure is helpful..."

2. Name what doesn't
"...but the tone is too formal."

3. Give one direction
"Rewrite this in a warmer, more conversational voice."

This keeps the system oriented without overwhelming it.

Why correction builds trust

Trust does not come from perfection.
It comes from recoverability.

When you know you can correct misalignment quickly, you stop fearing mistakes.

That confidence changes how you work:

- you experiment more
- you start faster
- you judge instead of hope

AI becomes less intimidating because you are no longer dependent on it being right the first time.

Misalignment as progress

Every correction teaches alignment.

Every adjustment reduces future friction.

Over time:

- fewer restarts
- fewer rewrites
- fewer "almost but not quite" moments

You are not fixing mistakes.
You are building a shared understanding.

That is progress.

Practice:
Try This Now (Chapter 9)

Purpose:
Learn to correct AI without restarting or lowering standards.

This practice trains precision instead of frustration.

What to Do: Use a recent AI output that felt "close but wrong."

If you don't have one, ask AI to write a short paragraph on any topic.

Step 1 — Identify the Issue
Before responding, decide:

- What specifically feels off?
- Choose one issue only.

Step 2 — Apply Targeted Correction
Reply with a short directive like one of these:

- "Keep the structure but make the tone warmer."
- "This is too detailed – simplify."
- "You focused on the wrong priority — emphasize clarity instead."

Step 3 — Observe the Adjustment
Notice whether the system:

- preserved what worked
- adjusted only what you named
- avoided introducing new problems

Reflection Questions:
Answer privately:

- Did this feel easier than restarting?
- Did the correction sharpen your own judgment?
- Did the second version feel closer to your intent?

Calibration Step:
If AI overcorrects, reply:

"That's too much change. Adjust only what I mentioned."

Quick Win:
You've learned how to turn misalignment into momentum instead of frustration.

CHAPTER 10

Your Leadership Advantage in the Age of AI

Leadership has never been about having all the answers.

It has always been about creating clarity when things are uncertain, helping others think when pressure is high, and holding standards without losing humanity.

AI does not replace those responsibilities.
It amplifies them — if you use it correctly.

This chapter is not about "AI leadership trends."
It is about how thoughtful leaders can use AI to lead better, not louder.

The quiet advantage most leaders overlook

Many leaders feel overwhelmed not because they lack skill, but because they lack **space.**

Space to:

- think before responding
- slow down complex decisions
- choose words carefully
- test ideas without consequence

AI provides that space.

Not by making decisions for you —
but by absorbing some of the cognitive load that drains judgment

When leaders use AI well, they gain:

- calmer decision-making

- clearer communication
- more consistent reasoning
- less reactive behavior

These are leadership advantages — not technical ones.

AI as a leadership thinking room

Think of AI as a private thinking room.

A place where you can:

- draft a difficult message before sending it
- reason through a decision before announcing it
- test how something will land
- explore options without committing

This matters because leadership often happens in public, under scrutiny.

AI gives you a place to think **without an audience.**

That alone changes how you show up.

Supporting leadership judgment, not replacing it

A critical boundary must remain clear:

AI does not carry responsibility.
You do.

AI can surface options.
It can outline tradeoffs.
It can reframe problems.

But it cannot weigh:

- trust
- culture
- history
- human impact

Those require judgment.

Strong leaders use AI to support judgment, not outsource it.
Weak leaders look for certainty where none exists.

Leading through complexity with clarity

Modern leadership is rarely about simple choices.

It involves:

- competing priorities
- incomplete information
- emotional stakes
- long-term consequences

AI helps by:

- organizing complexity
- slowing reasoning
- making tradeoffs visible

This allows leaders to respond with intention instead of urgency.

Urgency creates noise.
Clarity creates trust.

Using AI to improve communication

One of the most practical leadership uses of AI is communication.

Leaders often know what they want to say but struggle with:

- tone
- framing
- clarity
- brevity

AI helps you:

- draft messages calmly
- remove unnecessary edge
- adjust tone without weakening content
- align words with values

This is especially useful when emotions are involved.

AI lets you pause.

And pausing is often the difference between leadership and reaction.

The ethical responsibility of AI-assisted leadership

With this advantage comes responsibility.

Leaders must be intentional about:

- transparency
- authenticity
- accountability

AI should support your voice — not replace it.

If AI becomes a shield that distances you from people, trust erodes.

If AI becomes a tool that helps you communicate more thoughtfully, trust grows.

The difference lies in how you use it.

Leadership is still human work

No matter how capable AI becomes, leadership remains human work.

It requires:

- empathy
- courage
- integrity
- consistency

AI does not diminish these qualities.
It gives you a place to practice them more deliberately.

Used well, AI does not make leaders colder.

It makes them steadier.

Practice:
Try This Now (Chapter 10)

Purpose:
Use AI as a leadership thinking partner — not a decision-maker.

This practice helps you slow down and reason through a real leadership situation.

What to Do:
Think of a leadership decision or communication you're facing. It can be small.

Open a new AI conversation.

Prompt to Use:

"Help me think through a leadership situation. Ask clarifying questions first. Then help me reason through options, risks, and values-based considerations before drafting anything."

Step-by-Step:

1. Briefly describe the situation.
2. Answer the clarifying questions AI asks.
3. Review the options it presents.
4. Decide what you believe is the right path.

Reflection Questions:
Answer privately:

- Did this reduce emotional reactivity?
- Did it help you see tradeoffs more clearly?

- Did it support your judgment rather than override it?

Calibration Step:
If the response feels too clinical, reply:

Factor in emotional impact and long-term trust."

Quick Win:
You've used AI to create space — and space is a leadership advantage.

CHAPTER 11

The Real Transformation

Most people expect the transformation from using AI to be external.

They expect:

- faster output
- better drafts
- increased productivity
- cleaner execution

Those things do happen.

But they are not the real transformation.

The real change happens internally — in how you think, how you approach problems, and how you relate to your own uncertainty.

The shift you don't notice at first

At some point, often quietly, you begin to notice that starting is easier.

The blank page is less intimidating.
Complex problems feel more approachable.
You no longer wait for perfect clarity before engaging.

That is not because AI removed the difficulty.

It's because AI removed the friction.

Friction drains energy faster than effort ever did.

When friction is reduced, you regain momentum.

How support changes cognition

When you have a thinking partner available:

- you hold fewer variables in your head
- you externalize uncertainty instead of suppressing it
- you reason through problems instead of circling them

This changes cognition.

Your brain stops bracing against uncertainty and starts engaging with it.

You are no longer alone with the problem — and that matters more than most people realize.

The difference between dependence and support

Some people worry that working with AI will make them dependent.

That fear misunderstands what is actually happening.

Dependence replaces capacity.
Support restores it.

When AI is used correctly, you are not thinking less.
You are thinking better.

You are not handing off responsibility.
You are creating conditions for clearer judgment.

If AI disappeared tomorrow, you would still retain:

- the questioning habits
- the clarity frameworks
- the reasoning process
- the confidence to start imperfectly

Those are human skills.

AI helped you practice them — it did not replace them.

Why confidence feels different now

The confidence that emerges from this work is not bravado.

It doesn't announce itself.

It shows up as:

- quicker starts
- fewer restarts
- calmer revisions
- less self-criticism

You no longer need to prove intelligence.

You trust your ability to refine.

That trust changes how you show up everywhere — not just in AI conversations.

Transformation through repetition, not insight

This change does not happen because of one powerful insight.

It happens because you:

- asked questions repeatedly
- corrected misalignment
- refined thinking in small steps
- stayed in the loop

Over time, your nervous system learns that uncertainty is workable.

That is a deep transformation.

When you realize something has changed

Many readers reach a moment where they think:

"I'm not afraid of starting anymore."

That moment matters.

It signals that you no longer equate uncertainty with failure.

You see it as part of the process.

That is what this book was designed to teach.

Transformation that extends beyond AI

These changes do not stay contained.

They show up when:

- you write emails
- you lead meetings
- you make decisions
- you face ambiguity

AI was the training ground.

The skill is transferable.

That is the real return.

Practice:
Try This Now (Chapter 11)

Purpose:
Notice internal changes that are easy to miss but critical to long-term growth.

This practice helps you recognize progress that is not measured in output.

What to Do:
Open a new AI conversation or continue an existing one.

Prompt to Use:

"Reflect back to me how my thinking and interaction style have changed over the course of our conversations. Focus on patterns, not praise."

After You Receive the Response:
Read it slowly.

Notice what feels accurate and what surprises you.

Reflection Questions:
Answer privately:

- Where do I feel more confident than before?
- Where do I hesitate less?
- Where have I stopped overperforming?

Calibration Step:
If the response feels generic, reply:

"Base this only on how I have actually interacted with you, not on assumptions."

Quick Win:
You recognize growth that builds confidence without pressure.

CHAPTER 12

A Different Kind of Confidence

The confidence this work builds is not loud.

It does not announce itself with certainty or bravado.
It does not require you to be right the first time.
It does not come from mastery in the traditional sense.

It comes from **process.**

When you know how to begin, how to refine, and how to recover from misalignment, fear loses its grip.

You stop waiting to feel ready.
You stop needing perfect clarity before acting.
You stop interpreting uncertainty as incompetence.

That is a different kind of confidence.

Confidence rooted in repeatability

Traditional confidence often depends on outcomes.

If things go well, confidence rises.
If things go poorly, confidence collapses.

Process-based confidence is different.

Process-based confidence is different.

It comes from knowing:

- how to start when things are unclear
- how to ask for help without shame

- how to correct course without panic
- how to refine instead of restart

You are no longer betting your self-worth on the first attempt.

You trust the loop.

Why this confidence lasts

Tools change.
Technologies evolve.
Platforms disappear.

Process endures. The skills you have practiced in this book are not tied to AI. They are thinking skills:

- clarifying intent
- articulating priorities
- responding to feedback
- iterating without self-judgment
- holding authority while inviting collaboration

AI simply gave you a place to practice them consistently.

That means this confidence does not disappear if the tool changes.

It stays with you.

When confidence shows up unexpectedly

Many people don't notice this confidence at first.

It shows up quietly when:

- you open a blank document without hesitation
- you start a difficult conversation with more calm
- you revise instead of scrap
- you say, "Let me think this through," instead of rushing

Those moments are evidence of change.
They signal that you no longer confuse speed with competence.

Confidence without performance

Perhaps the most important shift is this:

You stop performing intelligence.

You no longer feel the need to:

- sound impressive
- hide uncertainty
- prove capability
- justify curiosity

You ask what you need to ask.
You take the time you need to take.

That freedom is not laziness.
It is maturity.

Carrying this forward intentionally

As you continue working with AI, remember what created this confidence:

- honesty over polish
- clarity over complexity
- correction over restart
- partnership over control

When something feels off, name it.
When you are unsure, say so.
When alignment drifts, correct it.

These habits compound.

What it means to "think like you"

Teaching AI to think like you was never about copying your words.

It was about reflecting your **judgment**.

Your values.
Your standards.
Your way of deciding what matters.

Now you have a process for doing that — deliberately, repeatably, and calmly.

That is confidence worth keeping.

Practice:
Try This Now (Chapter 12)

Purpose:
Anchor confidence in a repeatable thinking process you can use anywhere.

This practice turns what you've learned into a portable framework.

What to Do:
Open a new AI conversation.

Prompt to Use:

"Summarize the step-by-step process I now use to approach complex or unclear problems, based on how I've been working with you."

After You Receive the Response:
Read the steps carefully.

Notice which ones feel accurate and which need adjustment.

Refinement Step:
Reply with:

"Adjust this so it reflects how I actually pause, clarify intent, consider options, and then decide."

Reflection Questions:
Answer privately:

- Can I apply this process without AI?
- Do I trust this sequence when things feel messy?
- Where has this already helped me outside this book?

Quick Win:
You leave with a mental model you can reuse in any situation that requires clear thinking.

FINAL REFLECTION: A LETTER TO YOU

You did not come to this book looking for shortcuts.

You came because something mattered enough to think about carefully.

You wanted:

- clarity instead of noise
- support without surrender
- help without loss of authority

That intention matters.

This book was never about teaching you what to think.
It was about helping you **hear your own thinking more clearly.**

AI was not the answer.

AI was the partner that made space for the answer to emerge.

If there is one thing to remember as you move forward, let it be this:
You do not need certainty to begin.
You do not need perfection to progress.
You do not need permission to think deeply.

You already know how to lead.
You already know how to judge.
You already know how to correct course.

Now you also know how to work with a partner that reflects those strengths instead of dulling them.

Whenever you face something complex, unclear, or emotionally charged, you know what to do.

You pause.
You clarify.
You ask.
You refine.
You decide.

That process is yours now.

And it will carry you far beyond this book.

ABOUT THE AUTHOR

Christopher Pyle is a leadership coach, writer, and founder of Pyle of Goodness Press. His work focuses on clarity, kindness, and practical leadership in complex modern environments. Christopher writes to help people think more clearly, lead more humanely, and build systems that reflect their values. He believes technology should amplify humanity, not replace it.

Made in the USA
Monee, IL
04 November 2026

67381200R00066